Fact #1

Fruits get their vibrant colors from natural chemicals called pigments.

Fact #2

The sour taste of citrus fruits like lemons and oranges comes from citric acid.

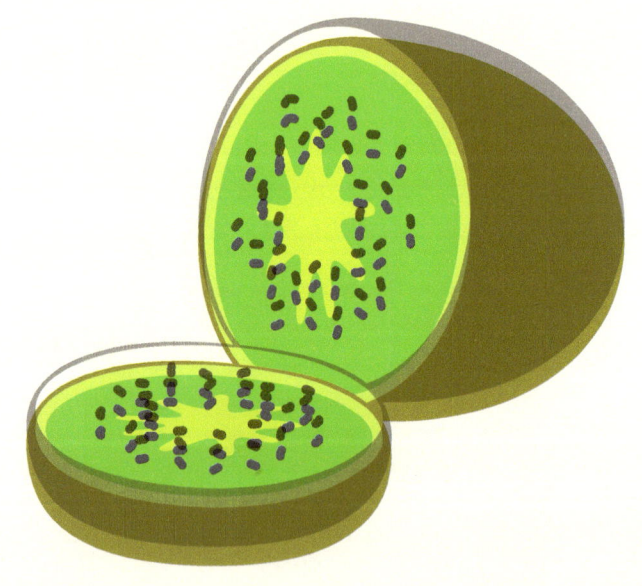

Fact #3

Apples and pears contain a chemical compound called ethylene, which helps them ripen.

Fact #4

Bananas release ethylene gas as they ripen, which can cause other fruits nearby to ripen faster.

Fact #5

The sweet taste of fruits comes from natural sugars like fructose and sucrose.

Fact #6

Many fruits contain antioxidants, which help protect our cells from damage.

Fact #7

The chemical compound responsible for the spicy taste in peppers is called capsaicin.

Fact #8

Avocados contain healthy fats called monounsaturated fats, which are good for the heart.

Fact #9

Blueberries contain a compound called anthocyanin, which gives them their blue color and has antioxidant properties.

Fact #10

Pineapples contain an enzyme called bromelain, which can help with digestion.

Fact #11

The tart taste of cranberries comes from organic acids like citric acid and malic acid.

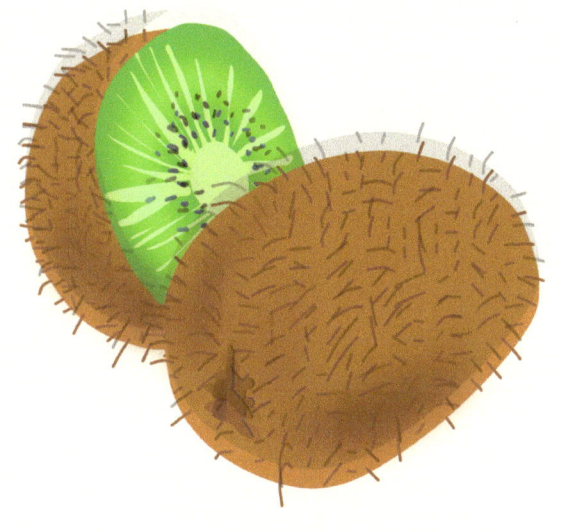

Fact #12

Oranges contain vitamin C, also known as ascorbic acid, which is essential for a healthy immune system.

Fact #13

The distinctive aroma of strawberries comes from a compound called methyl anthranilate.

Fact #14

Many fruits, like grapes and cherries, contain natural sugars that provide energy for the body.

LET'S RECAP

What gives fruits their vibrant colors?

Pigments

What compound gives citrus fruits their sour taste?

Citric acid

What chemical helps apples and pears ripen?

Ethylene

Which gas do bananas release to ripen faster?

Ethylene

What gives fruits their sweet taste?

Natural sugars

What do antioxidants in fruits help protect our cells from?

Damage

What compound causes the spicy taste in peppers?

Capsaicin

What kind of fats are found in avocados?

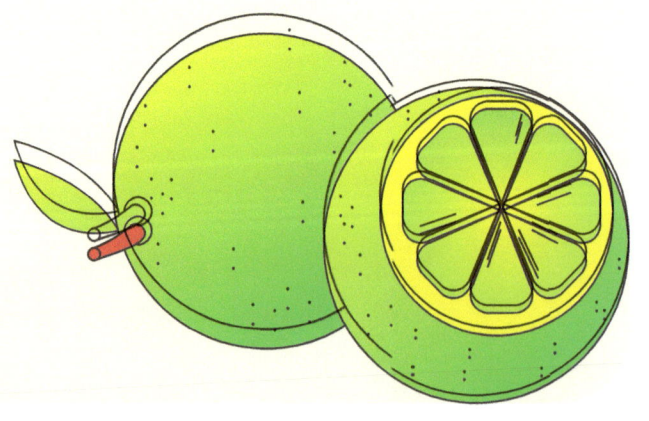

Monounsaturated fats

What gives blueberries their blue color and antioxidant properties?

Anthocyanin

What enzyme found in pineapples aids in digestion?

Bromelain

Which acids give cranberries their tart taste?

Citric acid and malic acid

What essential vitamin is found in oranges?

Vitamin C

What compound contributes to the aroma of strawberries?

Methyl anthranilate

What do natural sugars in fruits provide for the body?

Energy

ABOUT THE AUTHOR

Nisha is an educational professional with a fervor for storytelling and a background in science. She loves storytelling in her classrooms and loves to work on the social-emotional learning of young minds. She loves to create helpful content for learning and reading. She believes in making this world a better place by sensitizing people with better teaching-learning-knowing processes. should you have any suggestions, write us at nishawrites18@gmail.com. Your feedback and suggestions are important to us. Happy reading.

www.ingramcontent.com/pod-product-compliance
Lightning Source LLC
Chambersburg PA
CBHW051929210526
45473CB00006B/2186